SpringerBriefs in Water Science and Technology

For further volumes:
http://www.springer.com/series/11214

Pradipta Kumar Deb

An Introduction to Mine Hydrogeology

 Springer

Pradipta Kumar Deb
MGMI
Kolkata, West Bengal
India

ISSN 2194-7244   ISSN 2194-7252   (electronic)
ISBN 978-3-319-02987-0   ISBN 978-3-319-02988-7   (eBook)
DOI 10.1007/978-3-319-02988-7
Springer Cham Heidelberg New York Dordrecht London

Library of Congress Control Number: 2013951769

To my beloved wife Sikha for everything...........

Preface

Over three and a half decades of professional involvement in the field of hydrogeology and mining has necessitated in writing a handbook on Mine Hydrogeology which would serve the mining professionals, mine hydrogeologists, and graduate students. Keeping this requirement in mind, an attempt has been made to write this book which includes original field research of the author for various mine hydrogeological studies.

All original sources are acknowledged wherever are appropriate. References have been given at the end of each chapter.

My efforts would be successful if this book comes in handy to those for whom it is written.

P. K. Deb

Acknowledgments

In writing this book immense help has been taken from the following books:

(1) Groundwater by H. M. Raghunath.
(2) Groundwater Resource Evaluation by William C. Walton.
(3) Environment Impact Assessment Guidance Manual for Mining of Minerals by Ministry of Environment and Forests, Government of India.

The author deeply acknowledges their help in this matter. Computerization of this manuscript has been done by Soumyadipta Deb whose contribution has been deeply acknowledged.

Lastly, without the constant support and encouragement of my family members especially my son Suryadipta Deb at Virginia Tech University, USA this book could not have been written. My heartfelt deep acknowledgment goes to them.

P. K. Deb

Contents

Abstract

Very often, in India, a mine planner or a mining engineer faces groundwater seepage problems, in either in a running mine or in an open mine (whether open pit or underground). This calls for a thorough in-depth hydrogeological analysis of the mining area. In doing so, a professional mine hydrogeologist has threefold responsibilities: (a) to ensure water supply to the mine or colonies or process plants, (b) to assess the groundwater seepage potential into the mines, and (c) to assess environmental impact on the surrounding hydrogeological regime.

In this book, various case studies in relation to mine hydrogeological studies have been discussed. This study is unique because it emphasizes on the physical impact of mining on the hydrogeological environs, which has been dealt with a number of Indian case studies. This type of study has not been carried out earlier.

Mine hydrogeology deals with the hydrogeological study of a mine, opencast or underground, in order to fulfil various requirements of the mine. In doing so, water well inventory in the area should be conducted and a water table map or an isopiestic map of the area should be prepared. Hydrogeological investigation needs to be carried out along with groundwater exploration through well drilling. Pumping tests of the exploratory well/s need to be carried out in the form of Step Drawdown Test (SDT) and Aquifer Performance Test (APT). These would help to determine the discharge, drawdown, and aquifer parameters like T, S, and K. Depending on the water demand of mines, colonies and process plants and based on the pumping test data, planning for groundwater development for the area needs to be proposed.

For assessing groundwater seepage in the mine, mine development planning has to be understood first. The optimum pit depth and pit limit have to be worked out or known from the mine planners and aquifer disposition in the mine area needs to be depicted. Basic objective is to keep the water table below the working bench level in case of opencast mine, so that mineral or coal can be excavated without any seepage problem. Therefore, groundwater seepage computation has to be associated with the mine development planning. An underground mining project is initiated by driving an incline or adit or by lowering a shaft cutting through the geological formations. This means puncturing of both unconfined and confined aquifers which contributes toward groundwater seepage to the mine. Proper hydrogeological studies are required in such cases.

Environmental impact assessment study of mining on hydrogeological regime is statutory and needs to be carried out during the course of a running mine.

Keywords Hydrogeology · Mine hydrogeology · Mine development planning · Environmental impact · Aquifer parameters

Chapter 1
Introduction

1.1 World Ground Water Availability

Groundwater is one of the most important renewable resources of the Earth and it occurs in many sites. Water in the form of rainfall and river or stream water penetrates the ground and moves through the pores of soils, rocks, and rock fractures and joints which are permeable enough to transmit water through them. The Earth's total water resources have been estimated to be of the order of 1.37×10^8 million ha-m. Out of these water resources, about 97.2 % is salt water, available only in the Oceans and only 2.8 % is available as fresh water at any given time on our planet. Out of these 2.8 about 2.2 % is found as surface water and rest only 0.6 % as ground water. It is interesting to note that out of 2.2 % surface water, 2.15 % fresh water locked in the glaciers and ice caps and only to the tune of 0.01 % (i.e. 1.36×10^4 M h am) is available in reservoirs and lakes and only 0.0001 % in streams, 0.001 % as water vapour in atmosphere and 0.002 % as soil moisture in the 0.6 m of top soil. Out of 0.6 % ground water which is stored in the sub surface, only about 50 % of it (i.e. 41.1×10^4 M ha-m) can be economically exploited which is occurring down to a depth of 800 m. Presently about one-fifth of all global water being used is groundwater (Raghunath 1982, 1987).

1.2 Indian Ground Water Scenario

India receives average annual rainfall to the order of 1,140 mm. Based on this Central Government had estimated that the total annual rainfall in India is 370 M ha-m and one-third of it is lost in evapotranspiration. Out of rest 247 M ha-m of water, 167 M ha-m flows as run off and the balance 80 M ha-m trickles down as sub soil water. Out of this 80 M ha-m, 43 M ha-m gets absorbed as soil moisture in the layer whereas the remaining 37 M ha-m is recharge to the ground water from rainfall. The total annual ground water recharge from rainfall

P. K. Deb, *An Introduction to Mine Hydrogeology*,
SpringerBriefs in Water Science and Technology,
DOI: 10.1007/978-3-319-02988-7_1, © The Author(s) 2014

and seepages from canals and irrigation systems have been worked out to be 67 M ha-m (Raghunath 1982, 1987). Due to rapid urbanization and development ground water is being withdrawn heavily at many places which exceeds the rate of replenishment. This has caused some ecological problems like fluoride contamination in some hard rock areas like Purulia and Bankura districts of West Bengal and also problems of arsenic contamination and land subsidence in some alluvial areas of West Bengal (News items in local news papers).

1.3 Ground Water Condition at Some Indian Mines

Very often, in India, a mine planner or a mining engineer faces ground water seepage problem either in a running mine or in opening a mine (whether open pit or underground) which calls for a through in-depth hydrogeological study of the mining area. In doing so, a professional mine hydrogeologist has three fold responsibilities: (a) to ensure water supply to the mine or colonies or process plants, (b) to assess the ground water seepage dimension into the mines and (c) to assess environmental impact on the surrounding hydrological regime.

Ground water problems do not occur when mining is carried out in the hilly region. But when mining is continuous below ground surface and touches the water table of the region, ground water seepage problems begin and generally, the condition worsens when the mine goes deeper. In India, in most of such cases, the mining engineer resorts to pumping without proper hydrogeological studies causing continuous inundation of the mine leaving behind precious mineral resources. Such cases have been observed in most of the limestone and coal mines although such cases have been noticed in some of manganese mines also.

1.3.1 Sonadih Limestone Mines

In limestone mining areas, where 'karstification' has taken place during Geological Time, solution cavities have been developed through which tremendous ground water flow takes place. One such case was Sonadih limestone mine in Raipur district of Chhattisgarh State (previously it was Madhya Pradesh—M.P.).

The lime stone mining and cement plant lease area in Sonadih village was 4.4 sq. km. It was a green field project for which hydrogeological investigations were carried out keeping two clearly defined objectives in mind.

1. Assessing ground water potential for augmenting water supply to the proposed plant, mines and the township.
2. Assessing the dimension of ground water seepage problem in mining and planning for ground water seepage control.

It was estimated that in the initial phase, the water demand was 5,000 m³/day whereas this would increase to 8,000 m³/day in the final phase.

The area belongs to part of the famous Chattisgarh basin and is underlain by the rocks which are stratigraphically grouped as Raipur Series of Algonkian Age. The Raipur series is mainly composed of stromatolitic limestone, dolomite and intermittent bands or patches of shale. The limestone is overlain by top soil at places. The average thickness of the soil is about one meter with a variation from 0.20 m to more than 5 m at places. The top soil resembles Terrarosa which is developed in a typical karstic limestone terrain by the action of descending meteoric water. The limestone is stromatolitic and is near horizontal to low dipping—it dips 5°–6° towards NW with a general strike direction of NE–SW (Krishnan 1968).

Geological exploration through diamond core drilling has revealed the sub-surface geology of the area in much details. Although the limestone appears to be massive, hard and compact, in sub-surface in a number of places it has given rise to cavities/cavernous limestone.

The Chattisgarh basin is drained by two major river systems Godavari and Mahanadi rivers thereby making it Godavari-Mahanadi basin. Shivnath river being a major tributary to Mahanadi flows just outside the northern boundary of the project area. Therefore, the area may be classified as a part of Shivnath sub basin.

Groundwater occurs in the area both under unconfined and confined condition. The water level in the unconfined aquifer varies from 0.20 m to seven meters below ground level'. During monsoon, the water level varies from flowing condition (above ground level) to 6.10 m below ground level. Isobath map (i.e. water table map) constructed over the area shows that the ground water flow direction is towards Shivnath river. The equipotential lines crossing some of the nala section establishes the fact that drainages are mainly of influent nature (i.e. losing) towards southern and central part whereas it is of effluent nature towards north which makes Shivnath river a gaining one (Deb 1987).

Considering average top of the saturated unconfined aquifer as three meters below G.L., the total saturated thickness of the aquifer becomes 27 m down to 30 m below ground level. Within this saturated portion, two categories of ground water reserves would be available:

(a) Annual dynamic reserve
(b) Static reserve.

The annual dynamic reserve is the quantum of renewable annual ground water inflow within the zone of seasonal fluctuation in the unconfined or water table aquifer.

The static water reserve is the quantum of ground water available below the zone of seasonal fluctuation.

The annual dynamic reserve of the project area has been computed as 1.188×10^6 m³ whereas the static water reserve has been calculated to be 36.96×10^6 m³ (Walton 1970).

It was planned to mine the limestone deposit with six meter high benches with shovel dumper combination. As the cement grade limestone occurs primarily down to 24 m depth, bench wise ground water seepage along with water flow due to rainfall were computed and accordingly advise was given to the mine planners in advance. In order to facilitate proper dewatering, it was recommended to prepare a sump at a suitable location on the bench floor and 2–3 % inclination in the pit floor alignment was required towards the direction of ground water flow. The details of this project has been discussed in Chap. 7.

1.3.2 Limestone Mining and Cement Plant Project at Nongkhleih in Jaintia Hills District, Meghalaya

Towards establishing the feasibility of setting up the proposed 0.75 million tones per annum capacity cement plant along with limestone mine at Nongkhleih in Meghalaya, a large multi national company commissioned a ground water exploration study to establish the water availability and ground water conditions in the area, as the water requirement for the proposed cement plant would be about 1,000 m^3/day.

The area forms a part of submontane valley characterized by the presence of a rolling topography with high limestone ridges on the east and north east. Drainages are comparatively scanty and the drainage pattern is of parallel to sub parallel type indicating that it is controlled by structures.

The area is underlain by semi consolidated to consolidated sedimentary rocks of Shella formation falling under Jaintia Group of Tertiary age. The major litho-units are sandstone, shale and limestone. The area exposes the contact zone between sandstone and limestone of Shella formation with intercalated layers of shale (Dey 1968). Ground water exploration was carried out with five numbers. Exploratory tubewells and one observation well which were drilled with a Down the hole hammer (DTH) drill rig. Aquifer thickness varied from 18.36 to 44.20 m in the wells. Ground water occurs in the valley area under semi-confined to confined condition and depth to water varied from 4.10 to 12.56 m below ground level. The flow from the wells varied from 13,702 m^3/h to as high as 64.286 m^3/h with drawdown below two meters. Pumping tests were carried out on all the wells and co-efficient of transmissibility values varied from 61 to 285.33 m^2/day which ensures good amount of ground water might be withdrawn from the wells. Ground water exploration had confirmed that the water demand of 1,000 m^3/day would be easily met from groundwater regime of the area and one million USD which was planned to invest for the proposed cement plant could be shown green signal (Deb 2001). Details of this project would be discussed in Chap. 7.

1.3.3 Kathautia Open Cast Coal Mine, District Palamau, Jharkhand

In this open cast coal mine of Daltonganj coalfield, Gondwana coal occurs with sandstone and shale beds which is overlain by seven meters thick sandy loam and sand. Three benches of nine meters height have been developed where coal occurs in the lower benches. Coal is being mined with shovel dumper combination. Heavy ground water seepages have been noticed from the top bench which frequently hinders mining activity. The geological section of the mine is as under:

Ground level–7.00 m: Sandy loam, Sand
7.00–15.00 m: Sandstone
15.00–18.00 m: Coal seam

It was informed by the mine authorities that the north western mining face would be developed in the future whose present length is about 200 m.

From a careful study of the mine face vis-à-vis groundwater seepage, it was proposed that 6 inch a borewells with perforated casing (down to 15 m) should be drilled on the top bench down to a depth of 75 m, which is supposed to be ultimate pit depth. Ten such wells should be drilled at a distance interval of 20 m from each other so that during pumping, cone of depression formed in each well interferes with each other in order to lower the pumping water below the coal bench. This would ensure uninterrupted coal mining. As the mine goes deeper, the rate of pumping should be further increased in order to keep the water table below mining bench.

1.3.4 Neyveli Lignite Mine, Tamil Nadu

It was a unique case of artesian pressure for the aquifer below the lignite deposit. While planning for open cast mine, it was seen that the critical depth was 42.68 m below which it was dangerous to mine unless the aquifer pressure was reduced to below the bottom of lignite bed. The static water level of the aquifer was 30 m above mean sea level. The Lignite deposit occurring at Neyveli on top of the first aquifer separated by a thin layer of clay 1.5–3 m in thickness. Pioneering work was done by the Geological survey of India in collaboration with the Neyveli lignite Corporation in the 1960s. Ground water control operation of Neyveli Lignite Corporation started on July, 1961. A detailed account of this exercise would be described in Chap. 7 (Raghunath 1982, 1987).

1.4 Hydrogeological Investigations

Hydrogeology deals with the geological environs under which ground water is stored and moves through the saturated rock formation or group of formation. An aquifer is a saturated bed or rock formation which yields appreciable amount of water to be considered to be a potential source of water supply. Hydrogeological investigations are generally, taken up for water supply to various units viz industry, irrigation, domestic, townships etc. However, hydrogeological investigations are taken up for mining industry for three different reasons:

(a) To supply water for the mines, processing plant and mining colonies.
(b) To assess the dimension of ground water seepage into the mines and finding ways to control it by way of dewatering.
(c) To find out both quantitative and qualitative impact of mining on groundwater regime of the area.

Some of the mines are situated near a river or stream or streamlets. During monsoon, some of these rivers or streams flood the mine which creates problem in excavation. Besides, when groundwater is being dewatered from the mine nearby stream contributes water to the mine through permeable rock beds which stabilizes the water level in the mine. Therefore, proper pumping tests should be carried out in advance to establish the hydro-dynamic relationship between surface water and ground water and this needs to be quantified which would help in proper dewatering planning.

1.5 Environment Impact

When excavation is done in a mine, as soon as the water table is punctured, groundwater in rush takes place. In order to get rid of such situation, dewatering of the mine is carried out. This would have two effects on the mine surrounding:

(a) Lowering of water table in the nearby vicinity causing drying of wells in the villages.
(b) Contamination of ground water and nearby surface water.

In mine hydrogeological studies, such situations are frequently encountered, which needs to be properly addressed. This can be addressed by creating artificial recharge zones from where rainwater or stream water penetrates the ground surface and joins the groundwater table.This helps in replenishment of water table of the mining area.

References

Deb PK (1987) Report on the hydrogeological investigations in the leasehold area of the cement plant project. Sonadih, Raipur district., M.P (Phase-I), Tata Steel

Deb PK (2001) Hydrogeological investigation for proposed Cement Plant at Nongkleih, Jaintia Hills district, Meghalaya. Final report, Lafarge India Limited

Dey AK (1968) Geology of India. National Book Trust of India, New Delhi

Krishnan MS (1968) Geology of India and Burma. Higginbothams (P) India Ltd, Madras 7:194

Raghunath HM (1982, 1987) Groundwater. New Age International (P) Ltd., Madras 1:1–2, 10–13

Walton William C (1970) Groundwater Resource Evaluation. International Student Edition, McGraw-Hill Kogakusha Ltd., New Delhi 2:29–30

Chapter 2
Hydrogeology

The problems facing any ground water investigation programme are zones of occurrence and recharge. The various phases of a ground water investigation programme are as under:

1. Hydrometeorological study: collection of annual temperatures data, rainfall data, humidity data etc.
2. Hydrogeological study:

    (a) Geological mapping
    (b) Systematic water well inventory
    (c) Test drilling, sampling and logging
    (d) Pumping tests on wells (Aquifer tests).

3. Geophysical survey:

    (a) Surface
    (b) Down the hole.

4. Hydrogeochemical survey: Collection of ground water samples from inventoried wells and tubewells and preparation of hydrogeochemical maps on the basis of analytical data.
5. Computer modeling for groundwater basin management.
6. Water balance studies.
7. Conjunctive water resources management.

The objectives of any hydrogeological investigations are

1. Define recharge and discharge areas.
2. Define major water bearing units or aquifer.
3. Define location, extent and inter relationship of aquifers.
4. Establish physical parameters of aquifers like transmissibility, storage co-efficient and specific yield.
5. Establish hydrodynamic relationship between ground water and surface water, if any.
6. Estimate total sub surface storage capacity.

P. K. Deb, *An Introduction to Mine Hydrogeology*,
SpringerBriefs in Water Science and Technology,
DOI: 10.1007/978-3-319-02988-7_2, © The Author(s) 2014

7. Establish geological factors which affect quality of ground water.
8. Arrive at the location probable depth of drilling and yield from the borewell/
   tubewell (Raghunath 1982).

The hydrogeologist utilizes petrography, stratigraphy, structural geology and geomorphology in the search of ground water. Logs of wells and excavations either artificial or natural, give geological sections of the earth's crust and a study of the surface and subsurface distribution of rocks and their character, thickness and depth below land surface is prerequisite to an understanding of the occurrences and movement of groundwater at any locality. The hydrogeologist maps the aerial extent of the various lithologic units and their water yielding properties. In preparing hydrogeological maps, greater emphasis is placed on the description of aquifers and acquitards. The position and thickness and continuity of aquifers, aquitards and aquicludes are determined with stratigraphic tools (Walton 1970).

There are two types of aquifers: unconfined and confined. Unconfined aquifers are generally the 1st aquifer from the surface which are open to atmosphere. The water level in such unconfined aquifers has no pressure head and is called water table. In confined aquifers, aquifer is bounded on top and bottom by impervious rocks or sediments and the water level in the aquifer rises above the top of that aquifer. This water level is known as piezometric surface. Water level in wells are constantly fluctuating within a relatively short time. Water levels in wells in confined aquifers generally fluctuate to a greater extent than the water levels in water table aquifers.

From the water level data from various wells in a given area, Water table contour map (or iso both map) is constructed and flow net analysis is made. This throws light on the direction of ground water flow in the selected flow channels etc. Similarly, from water level data from confined aquifer wells, piezometric surface map (or isopiestic map) is constructed and flow net analysis indicates direction of ground water flow and along a particular cross section ground water inflow and outflow to and from the basin can be computed.

Hydrogelogical studies are very vital for the following sectors:

1. Water supply
2. Rural water supply
3. Mining
4. Civil construction
5. Road making
6. Dam and water reservoir
7. Ground water pollution.

This book deals with the hydrogeological studies related to mining. Mine hydrogeology deals with the hydrogeological study of a mine, opencast or underground, in order to fulfill various requirements of the mine.

Fig. 2.1 DTH rig in action

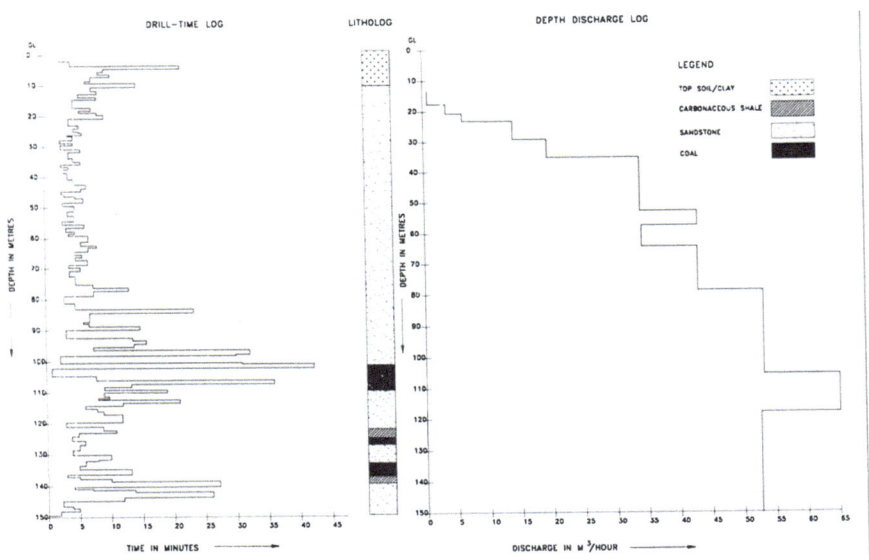

Fig. 2.2 Graphical representation of drill-time, lithological and depth discharge logs of exploratory tubewell ETW-II

For studying hydrogeological conditions of a mine area, ground water exploration is to be carried out. As most of the mine area in India is underlain by consolidated rocks/sediments, ground water exploration is carried out by deploying a DTH rig (i.e., Down the hole hammer rig) which drills the formations faster than any other type of rig (Fig. 2.1). Drill cutting samples are collected at every three metre interval or whenever there is change in formation. Litho-logical records along with drill time record and flow record (measured through a 90° V notch channel) are represented graphically which helps to identify the potential aquifer zines in the sub-surface. One such graphical example is shown in Fig. 2.2.

References

Raghunath HM (1982, 1987) Groundwater, vol 1. New Age International (P) Limited, Madras, pp 15–16

Walton WC (1970) Ground water, resource evaluation. International Student Edition, McGraw Hill Kogakusha Ltd., New Delhi, vol 2, pp 55–56

Chapter 3
Pumping Tests and Aquifer Parameters

Aquifer parameters which determine the characteristics of an aquifer are storage co-efficient (s), coefficient of transmissibility (T) and hydraulic conductivity or co-efficient of permeability (K). Storage co-efficient of an aquifer is the volume of water discharges from a unit prisom i.e. a vertical column of aquifer stranding on a unit area (1 m^2) as water level (piezometric level in confined aquifer) falls by a unit depth (1 m). For unconfined aquifers (water table aquifers) the storage co-efficient is the same as specific yield. The storage coefficient for confined aquifers ranges from 0.00005 to 0.005 and for unconfined aquifers from 0.05 to 0.30.

The coefficient of transmissibility (T) is the discharge through unit width of aquifer for the fully saturated depth under a unit hydraulic gradient and is usually expressed as lpd/m or m^2/day. It is the product of coefficient of permeability as (k) and saturated thickness of the aquifer (b). Thus $T = Kb$ and it has the dimension L^2/T.

The coefficient of permeability (K) is the ability of a formation to transmit water through its pores (whether primary or secondary) when subjected to a difference in head. It can be defined as the flow per unit cross sectional area of the formation when subjected to a unit hydraulic head per unit length of flow and has the dimension L/T. It is generally expressed as m/day.

3.1 Pumping Test

After the exploratary wells are drilled, pumping test on the wells are carried out. Two types of pumping tests are carried out on exploratory wells.

1. Step Drawdown Test (SDT)
2. Aquifer Performance Test (APT)

SDT is carried out at variable discharge in order to know the respective drawdown. The basic objective of such test is to measure the well performance so that an optimum yield at optimum drawdown can be arrived at. For groundwater

P. K. Deb, *An Introduction to Mine Hydrogeology*,
SpringerBriefs in Water Science and Technology,
DOI: 10.1007/978-3-319-02988-7_3, © The Author(s) 2014

control in mining projects, yield higher than the optimum yield is aimed at in order to achieve maximum drawdown. For a battery of wells, maximum yield from the wells will result in interference of drawdown which would ensure lowering of water table or piezometric surface below the mining benches so that uninterrupted mining can take place.

APT is carried out at a constant discharge for a relatively longer period of time in order to record drawdown at different times. The basic objective of such test is to determine the aquifer parameters as coefficient transmissibility (T), storage co-efficient (S) and coefficient of permeability (K).

Examples of SDT and APT which were carried out in Nongkhleih cement plant project, Meghalaya are being discussed in the following pages.

3.2 Pumping Test on ETW-II

Five numbers exploratory tubewells were drilled in the project. Exploratory tubewell (ETW-II) was drilled down to a depth of 150 m and an observation well (OBW-I) was also drilled at about 25 m north of ETW-II. Summarized lithological logs of ETW-II and OBW-I are given (vide Tables 3.1 and 3.2).

3.2.1 Step Drawdown Test

Before initiation of the test, statie water level was measured to be 4.25 m bmp. At the start the well was pumped at a discharge of 83 lpm and it was allowed to run

Table 3.1 Summarized lithological log of ETW-II

Lithology	Depth range (m)		Thickness (m)
	From	To	
Top soil, clay, yellowish brown, stiff, plastic and sticky	G. L.	10.43	10.43
Sandstone, medium grained, light yellowish grey, highly weathered with some pebbles and cobbles	10.43	34.82	24.39
Sandstone, light grey, fine to medium grained, highly fractured and jointed with vertical joints.	34.82	49.07	14.25
Sandstone, light grey, coarse, grained, highly fractured and jointed	49.07	57.62	8.55
Sand stone, light yellowish grey with alternation laminates of shale	57.62	84.63	27.01
Sand stone, light gray to greyish black, find to medium grained, with occasional laminates of coal, possiliferous at places	84.63	101.92	17.29
Coal grayish black, fractured with alternative bands of sandstone	101.92	109.97	8.05
Sandstone, light gray fine grained fractured and jointed with alternating laminaes of carboneceous shale and coal	109.97	150.00	40.03

Table 3.2 Summarized lithological log of OBW-I

Lithology	Depth range (m)		Thickness (m)
	From	To	
Top soil, clay, yellowish brown, stiff plastic and sticky	G. L.	10.43	10.43
Shale, mottled, with some pebbles and cobbles	10.43	13.48	3.05
Sandstone, medium grained, light yellowish grey, highly weathered	13.48	25.58	12.10
Sandstone, light grey, moderately hard, fractured and jointed	25.58	43.88	18.30
Sandstone grey fine to medium grained, fractured and jointed with alternating laminees of coal	43.88	68.45	24.57
Sandstone, light grey fine grained fossili ferrous, highly fractured and jointed with alternating thin laminaes of carb, shale and coal	68.45	88.50	20.05
Coal seam with inteermittents layers of carbonaceous shale and sandstone	88.50	97.80	9.30
Sand stone, light grey, fine grainted	97.80	102.45	4.65
Sand stone medium to coarse grained, weathered, fractured and jointed, with laminaes of coal	102.45	125.55	23.10
Sand stone light grey, fine to medium grained, hard and compact, fractured at places, with occasional thin laminaes of carbonaceous shale and coal	125.55	150.00	24.45

for 60 min. Discharge was gradually increased at the end of each hour and the test continued for 5 h. The summarized results of Step Drawdown Test (SDT) are as given in Table 3.3.

It may be observed that the well is of very high specific capacity and of almost uniform specific drawdown. Intererstingly, after step II, specific capacity of the well increased.

In order to attain the effective optimum yield of 38,500 m^3/h or 642 lpm of the well, the projected drawdown/would be around 2.64 m or say 2.75 m. Data in respect of SDT have been shown in Table 3.4. Graphical results of SDT has been depicted in Fig. 3.1.

Type of Pump	Submersible KSB with water-filled. Motor UMA 100-1, 3 phase
H.P	5 H.P
Dia	31/2''
Capacity	185 lpm
Diameter of Delivery pipe	11/2''
Date of test	21.07.2001
Static Water level	4.25 m b.m.p
M.P Top of well casing	0.15m agl

Table 3.3 Summarized results of SDT

Step	Duration (min)	Discharge (Q) lpm	Drawdown(s) in metres	Specific capacity (q/s) lpm/m	Specific drawdown (S/Q) m/lpm
I	60	83	0.40	207.50	0.005
II	60	103	0.55	187.30	0.005
III	60	160	0.64	250.00	0.004
IV	60	172	0.69	249.30	0.004
V	60	185	0.76	243.42	0.004

3.2.2 Aquifer Performance Test

Aquifer Performance Test (APT) was conducted on ETW-II for a continuous period of 480 min (i.e. 8 h) at a contrast discharge of 177 lpm. During the course of APT, regular monitoring of pumping water level was made in the observation well (OBW-I) while ETW-II was being pumped order to arrive at respective drawdown. At the end of the test, maximum drawdown in the observation well was recorded to be 0.964 m (vide Table 3.5).

Time drawdown data thus generated during the course of APT were plotted on a semi-logarithmic graph paper and a best fitting straight line was drawn (vide Fig. 3.2). This shows more or less non-equilibrium condition.

The modified non-equilibrium formula of Theis (1933) was used to compute co-efficient of Transmissibiltiy (T).

$$T = \frac{2.3\emptyset}{4\pi\Delta s}$$

where

T   co-efficient of transmissibility in m²/day

Q   Discharge in m³/day

Δs Change in meters in the drawdown over one log cycle of time.

T $\frac{2.3 \times 254.88}{4 \times 3.14 \times 0.22} = 212.15\,\mathrm{m^2/day}$

Hydraulic conductivity or permeability $K = \frac{T}{\Delta}$, where b is the aquifer thickness.

$$K = \frac{212.15}{44.20} = 4.80\,\mathrm{m/day}$$

The storage co-efficient, S, can be obtained by selecting any drawdown from the time drawdown curve for a given time and substituting this value in the following equation (Raghunath 1987).

$$S = \frac{2.3Q}{4\pi T} log_{10} \frac{2.25Tt}{\gamma^2 S}$$

Table 3.4 Step drawdown test on the ETW-II

Time in hours	Pumping water level in m (b.m.p)	Drawdown(s) in m	Discharge (Q) in l.p.m	Remarks
1,232	4.54	0.29	Step-1	Pumping started at 1,230 h
1,234	4.544	0.294	Q = 83	
1,236	4.54	0.29		
1,238	4.50	0.25		
1,240	4.46	0.21		
1,245	4.54	0.29		Atm temp = 250 °C
1,250	4.54	0.29		Water temp = 21.50 °C
1,255	4.53	0.28		
1,300	4.56	0.31		
1,310	4.566	0.316		
1,320	4.60	0.35	Q = 83	
1,330	4.65	0.40		
1,332	4.66	0.41	Step-II	
1,334	4.68	0.43	Q = 103	Water crystal clear
1,336	4.69	0.44		
1,338	4.72	0.47		
1,340	4.74	0.49		
1,345	4.765	0.515		
1,350	4.76	0.51		
1,355	4.78	0.53		
1,400	4.79	0.54		
1,410	4.79	0.54	Q = 103	
1,420	4.80	0.55		
1,430	4.80	0.55		
1,432	4.82	0.57	Step-III	
1,434	4.84	0.59	Q = 160	
1,436	4.85	0.60		
1,438	4.85	0.60		Water crystal clear
1,440	4.86	0.61		
1,445	4.84	0.59		
1,450	4.84	0.59		
1,455	4.85	0.60		
1,500	4.86	0.61		
1,510	4.88	0.63		
1,520	4.885	0.635		
1,530	4.89	0.64		
1,532	4.905	0.655	Step-IV	
1,534	4.90	0.65	Q = 172	
1,536	4.91	0.66		
1,538	4.91	0.66		
1,540	4.91	0.66		
1,545	4.91	0.66		
1,550	4.91	0.66		
1,600	4.925	0.675		

(continued)

Table 3.4 (continued)

Time in hours	Pumping water level in m (b.m.p)	Drawdown(s) in m	Discharge (Q) in l.p.m	Remarks
1,610	4.935	0.685		
1,620	4.94	0.69		
1,630	4.94	0.69	Q = 172	
1,632	4.95	0.70	Step-V	
1,634	4.95	0.70	Q = 185	
1,636	4.96	0.71		
1,638	4.97	0.72		
1,640	4.97	0.72		
1,645	4.98	0.73		
1,650	4.985	0.735		Water slightly dirty
1,655	4.98	0.73		
1,700	4.98	0.73		
1,710	4.98	0.73		Water crystal clear
1,720	4.99	0.74		Atm temp = 24.50 °C
1,730	5.01	0.76		Water temp = 21.50 °C
				Test concluded at 1,730 h

agl implies 'above ground level'
bmp implies 'below measuring point'

where

s	Drawdown in meters at any selected time on the Time-drawdown curve
Q	Discharge in m^3/day
T	Transmissibility in m^2/day
t	Any selected time on the Time-Drawdown curve, in days
r	Distance between pumped well and observation well, in meters
S	Storage co-efficient.

$$0.45 \quad \frac{2.3 \times 254.88}{4 \times 3.14 \times 212.15} \, log_{10} \frac{2.25 \times 212.15 \times 0.00069}{(25)^2 \times S}$$

or S 46.80×10^{-5}

The storage co-efficient value thus obtained clearly indicated that the aquifer is confined. The computed aquifer parameters indicated that the aquifers are very good to excellent with high yield potential especially when considered in the light of hard rock aquifers. Data in respect of APT has been shown in Table 3.5. During the course of pumping test, photograph was taken which has been shown in Fig. 3.3.

Date of test	22.07.2001 (of Observation Well OBW-I)
Static Water level	4.951 m bmp
M.P Top of well casing	0.41 m agl

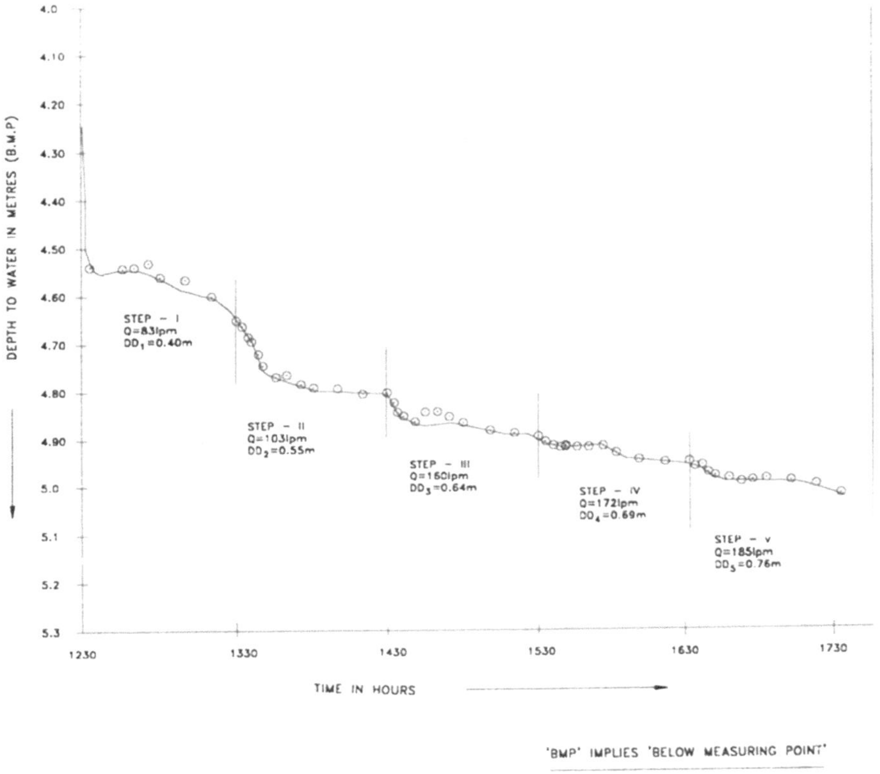

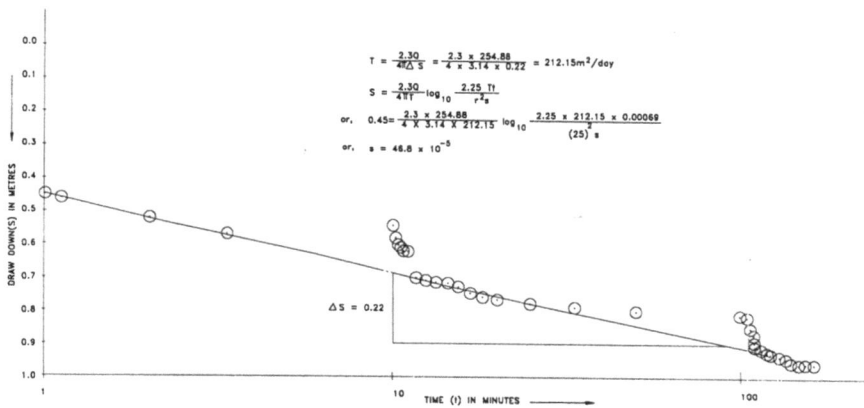

Fig. 3.1 Results of step drawdown test (SDT) on the exploratory well ETW-II

Fig. 3.2 Results of aquifer performance test (APT) on the exploratory well ETW-II (data collected from observation well OBW-I while ETW-II was being pumped out)

Table 3.5 Aquifer performance test on the exploratory well ETW-II

Time in hours	Time (t) since pumping started (min)	Pumping water level in m (b.m.p)	Drawdown(s) in (m)	Discharge(Q) in lpm	Remarks
901	1	5.401	0.45		Pumping started at 0900. Water level measurement on the observation well OBW-I which is situated at a distance of 25 m north of ETW-II
902	2	5.411	0.46	Q = 177	Q fluctuating, adjusted water sample
906	6	5.491	0.54		
908	8	5.521	0.57		ETW-II/S–I
910	10	5.490	0.539		
912	12	5.530	0.579		
914	14	5.545	0.594		Atm temp = 26 °C
916	16	5.565	0.614		Water temp = 21.8 °C
918	18	5.580	0.629		
920	20	5.585	0.634		
925	25	5.650	0.699		
930	30	5.655	0.704		
935	35	5.660	0.709		
940	40	5.670	0.719		
945	45	5.675	0.724		
950	50	5.695	0.744		
955	55	5.705	0.754		
1,000	60	5.710	0.759		
1,010	70	5.720	0.769		
1,020	80	5.734	0.783		
1,030	90	5.742	0.791		
1,040	100	5.755	0.804	Q = 177	
1,050	110	5.767	0.806		
1,100	120	5.775	0.814		
1,110	130	5.780	0.819		
1,120	140	5.790	0.839		
1,130	150	5.800	0.849		
1,140	160	5.820	0.869		
1,150	170	5.840	0.889		
1,200	180	5.845	0.894		Q fluctuating, adjusted water
1,220	200	5.850	0.899		sample
1,240	220	5.865	0.914		ETW-II/S–I
1,300	240	5.870	0.919		
1,320	260	5.875	0.924		
1,340	280	5.880	0.929		

(continued)

Table 3.5 (continued)

Time in hours	Time (t) since pumping started (min)	Pumping water level in m (b.m.p)	Drawdown(s) in (m)	Discharge(Q) in lpm	Remarks
1,400	300	5.885	0.934	Q = 177	Test concluded at 1700 h
1,430	330	5.890	0.939		
1,500	360	5.895	0.944		
1,530	390	5.910	0.959		
1,600	420	5.910	0.959		
1,630	450	5.915	0.964		
1,700	480	5.915	0.964		

agl implies 'above ground level'
bmp implies 'below measuring point'

Fig. 3.3 Pump test at Nongkleih, Meghalaya

References

Deb PK (2001) Hydrogeological investigation for proposed cement plant at Nongkhleih, Jainta Hills District, Meghalaya, Final Report, Lafarge India Limited

Kruseman GP, De Rider NA (1970) Analysis and evaluation of pumping test data, Inst. For land reclamation and improvement, Wageningen, The Netherlands

Raghunath HM (1982, 1987) Groundwater, New Age International (P) Ltd., Madras 5:135–203

Theis CV (1933) The relation between the lowering of the piezemetric surface and the rate and duration of discharge of a well using, groundwater storage. Trans Am Geophys Un 16:519–524

Walton WC (1967) Selected analytical methods for well and aquifer evalution, Bull 49, Illinois State Water Survey Division, Urbana

Chapter 4
Ground Water Quality

In hydrogelogical studies, groundwater quality assumes a very important role. The chemical, physical and bacterial characteristics determine the usefulness of groundwater in various sectors, viz domestic, municipal, industrial, agricultural and mining. Groundwater pollution requires special study on quality and remediation methods and process for this.

In mining areas, importance of groundwater quality study is of two fold: (a) providing safe drinking water to the mining colonies and nearby villages and (b) reducing toxic and harmful mine water discharge in order to avoid pollution of surface and groundwater in the peripheries.

In groundwater analysis constituents commonly determined are expressed as ions like cations including calcium, magnesium sodium and potassium and the anions including sulphate, chloride, fluoride, nitrate and those contributing to alkalinity which are usually expressed in terms of an equivalent amount of carbonate and bicarbonate. The acidity or alkalinity of water is generally expressed as pH which is concentration of hydrogen ion.

The water quality should satisfy the requirements or standards set for the specific use viz domestic, livestock, agricultural and industrial purposes.

The world Health organization (WHO) drinking water standards (1963) are as Table 4.1:

During the course of aquifer performance Tests on all the exploratory wells in Nongkhleih, Meghalaya, water samples were collected and analysed a few of them are displayed in the following Table 4.2.

It may be seen from Table 4.2 that pH of the water is almost neutral. The solids content and turbidity levels are very low. Iron is absent and coliform organisms were not detected in the samples. The water quality is thus very good and the parameters are well within the permissible limits for drinking purpose as stipulated in Indian standard IS: 10500 (Drinking Water Specification, IS 10500: 1991).

For ground water quality for mine water discharges and their remediation, the reader may refer suitable literature on Mine water (Younger et al. 2002).

P. K. Deb, *An Introduction to Mine Hydrogeology*,
SpringerBriefs in Water Science and Technology,
DOI: 10.1007/978-3-319-02988-7_4, © The Author(s) 2014

Table 4.1 WHO drinking water standards

Constituents	Limit of general acceptability (mg/l)	Allowable limit (mg/l)
Total dissolved solids (TDS)	500	1,500
Colours (^0H)	5	50
Turbidity	5	25
pH	7–8	Min 0.5
		Max 9.2
Chloride	200	600
Iron	03	1.0
Manganese	0.1	0.5
Copper	1.0	1.5
Zinc	5.0	15.0
Calcium	75.0	200
Magnesium	50.0	150
Magnesium and Sodium Sulphate	500	1,000
Nitrate (as NO_3)	45.0	–
Phenols	0.001	0.002
Synthetic detergents (ABS)	05	1.0
Caron-Chloroform extract	0.2	0.5

(Raghunath 1987)

Table 4.2 Ground water quality in Nongkheih

Sl. no.	Parameters and units	ETWII/S-1	ETWII/S-2	ETWV/S-5
1.	pH	7.15	6.9	6.95
2.	Conductivity at 25 °C (µmhos/cm)	70	100	130
3.	Turbidity (NTU)	<5	<5	<5
4.	Colour (^0H)	<5	<5	<5
5.	TDS (mg/l)	48.56	68.00	86.00
6.	Total coliform (MPN/100 ml)	Nil	Nil	Nil
7.	Total hardness (mg/l as $CaCo_3$)	60.00	48.0	64.00
8.	Calcium (mg/l as Ca)	17.64	17.60	22.40
9.	Magnesium (mg/l as Mg)	3.90	1.00	2.00
10.	Fluoride (mg/l as F)	BDL	BDL	BDL
11.	Chloride (mg/l as Cl)	8.93	10.00	8.00
12.	Sulfate (mg/l as So_4)	0.52	5.00	6.00
13.	Alkalinity (mg/l as $Caco_3$)	66.00	56.00	60.00
14.	Silica (mg/l as Sio_2)	0.20	0.10	BDL
15.	Sodium (mg/l as Na)	4.00	5.00	4.00
16.	Potassium (mg/l as K)	7.00	7.00	4.00
17.	Iron (mg/l as Fe)	BDL	BDL	BDL

BDL Below Detection level

References

Deb PK (2001) Hydrogeological investigation for proposed cement plant at Nonghkhleih, Jaintia Hills district, Meghalaya. Final report, Lafarge India Limited

Raghunath HM (1982, 1987) Ground water, vol 9. New Age International (P) Ltd, Madras, p 346

Walton WC (1970) Ground water resources evaluation, vol 7. International Student Editor. MCGraw Hill Kogakusha Ltd, pp 439–442

Younger PL, Banwart SA, Hedin RS (2002) Mine water: hydrology, pollution, remediation, vol 16. Springer Publishing

Chapter 5
Mine and Mine Development Planning

Primary step of mining of minerals is the removal of the deposits from the ground. Once the minerals/ore is removed, additional preparation process is required to isolate the valuable minerals from their waste gangue minerals. There are two basic method of mining of minerals: open cast and under ground mining. The choice of method depends on the geological, hydrogeological, geotechnical, geographic, economic, technological, environmental, safety, sociopolitical and financial considerations.

The major objective of mine development planning is to provide auxiliary and support facilities for physically opening a surface or underground mine and bringing it to full production. The surface facilities unique to underground mining are mine main entries (shafts, inclines and adits), head frames, heap, stead, storage bins, hoist houses etc., the additional underground facilities may consists of secondary and tertiary openings for providing access haulage and ventilation and various other facilities such as transportation crusher stations, power distribution equipments and numerous other installations.

In opencast mines, if the deposit does not outcrop, advanced stripping is required before the beginning of production. Development is carried out according to a carefully designed plan as worked out in the feasibility report.

5.1 Open Cast Mining

Surface or open cast mining is used for large, near surface deposits. Rock is drilled, blasted, loaded into dumpers and hauled to a place where it is crushed and ground to a uniform size for further processing. Open cast mining needs the removal and disposal of top soil and underlying rock known as overburden. Mining should be planned in such a manner so that it combines mining, processing and reclaiming the land simultaneously.

General geology and hydrogeology of the mine area and the geological sections (both transverse and longitudinal) in the prospect area are required to be prepared and studied. Natural and geological factors, topography, depth, geological

P. K. Deb, *An Introduction to Mine Hydrogeology*,
SpringerBriefs in Water Science and Technology,
DOI: 10.1007/978-3-319-02988-7_5, © The Author(s) 2014

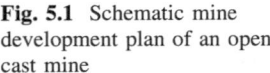

Fig. 5.1 Schematic mine development plan of an open cast mine

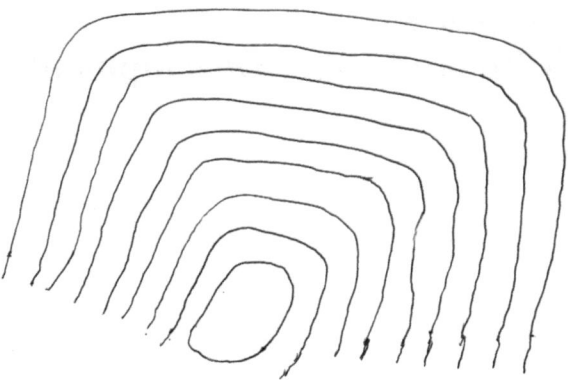

disturbances like faults, folds etc., climate, mineral/coal grade, mineable reserves within the mining lease boundary, production rate, estimated life of the mine need to be estimated before selection of the mine area. Depth of the deposit at the final pit limit, thickness and dip should be determined. Overburden thickness (maximum and minimum) is required to be estimated. Individual bench height and total no. of working benches are to be worked out (vide Fig. 5.1).

Maximum allowable stripping ratio, overall stripping ratio, mining plan (development and production) are to be critically studied. Year wise development and production plan for next 5 year needs to be worked out. All the mine plans are required to be approved by the concerned regularly authority.

5.2 Underground Mining

Underground mining methods are used where a deposit occurs beneath the Earth's surface. In order to reach the deposit for its removal, it is necessary to excavate either a vertical shaft, a horizontal adit or an inclined short. Within the ore/coal deposit, horizontal passage called drifts and crosscuts are developed on several levels to access mining areas to form stopes. Blasted rocks are hauled away from the stopes by loaders, rail or trucks which may bring these directly to the surface or transport these to a shaft where it is hoisted to the surface and sent to the crushing plant.

In both open cast and underground mining extraction of ore/coal and waste requires the use of heavy machinery and explosives. A variety of earth moving equipments such as backhoes, front end loaders tractors dumpers etc., are used to remove surface soil and soft waste rock. Small transportation equipments are used to transport the mineral/coal to the surface from under ground mines, large loaders and trucks are used to carry the material away from the mine site for further processing (Ministry of Environment and Forests, Government of India 2010).

Explosives such as mixtures of ammonium nitrate and fuel are used to blast away harder rocks.

Reference

Ministry of Environment and Forest, Government of India (2010) Environmental impact assessment guidance manual for mining of minerals, vol 1. pp 1–10

Chapter 6
Environmental Impact Analysis on Hydrological Regime

In India, Ministry of Forests and Environment, Government of India, has set certain guidelines for Mining Projects in order to study environmental impact on hydrological regime and their remediation, which are being discussed in the following few pages:

6.1 Anticipated Impact

Mining and its associated activities not only use a lot of water but also likely to affect the hydrological regime of the area. The major impact of deep and large mines (both underground and open cast) is of natural ground water table. Lowering of water table may result in reduced groundwater availability. Extraction of different minerals is known to lead to water pollution due to heavy metal, acid discharges and increased suspended solids. Deep underground mines directly affect the water table or piezometric surface of the area. However, the impact of mining project on hydrogeology and surface water hydrology are site specific and depends upon the characteristics of the mines hydrogeology and requirement of ground water for other uses.

The anticipated impact on hydrogeology of the mining area may be as under:

1. Regional surface and ground water movement.
2. Ground water inflow into the mine with subsequent contact with mining related pollutants'.
3. Surface water inflow and precipitation related recharge.
4. Increase in surface and ground water interaction with the mine working because of subsidence.
5. Loss of surface features such as lakes, streams and ponds through subsidence.
6. Pathways for closure flow resulting from adits shafts and overall mine design.
7. Operational and post closure geochemistry and resulting toxics mobility.

Specially, mine water, ground water withdrawal and land subsidence can potentially create environmental problems that cannot be easily corrected. The

P. K. Deb, *An Introduction to Mine Hydrogeology*,
SpringerBriefs in Water Science and Technology,
DOI: 10.1007/978-3-319-02988-7_6, © The Author(s) 2014

ground water recharge including ground water budgeting should be computed following standard methods.

6.2 Remediation Measures

Although water pollution in mining area is location and type of mining and minerals specific, mine water management calls for the following remediation measures:

1. The overall drainage planning should be made in such a manner that the existing pre- mining drainage condition should be maintained to the extent possible so that run off distribution is not affected.
2. Garland drains should be constructed on all side of quarries and external dumps. All the garland drains should be routed through adequately sized catch pits should be calculated on the basis of silt loading, slope, detention time required.
3. Retaining walls with weep holes should be built all round the external dumps. The storm water should pass through weep holes to the garland drains.
4. The surplus treated mine water should be discharged into local ponds agricultural fields which should act like a constant source of recharge to improve the ground water level in the area.
5. Rainwater harvesting by constructing check dams on natural nallah and developing water bodies should be planned for recharging groundwater.
6. Considering the composition, the mine drainage water, should be adequately treated before utilizing in agricultural fields in the surrounding through a planned network of drains pipe line.
7. The toxicity of the mine water should be treated specifically to meet the prescribed standard before discharge.
8. Settling tanks to treat mine water should be located preferably on a flat area, to reduce the risk of erosion.
9. Shallow and deep piezometers should be installed close to the mining area for long term monitoring of water levels in the aquifer/s.
10. Mine water management plans should be drawn at periodic intervals till closure of the mine (Ministry of Environment and Forests, Government of India 2010).

In India, physical and chemical impact of mining on hydrogeological regime in many mining areas are common. This not only lowered the water table of the nearby areas but also changed the chemical quality of groundwater. Because of this problem, there has been a stiff resistance from the local villagers to the running mining projects. Here lies the social responsibilities of the mining companies towards the local villagers to remediate the ill effects of groundwater pollution. As groundwater is a renewable resource, ample opportunities exist there to study and identify areas of groundwater recharge which can be protected and developed for direct infiltration of rainwater or streamwater into subsurface. This effort can resist the trend of lowering of watertable in the vicinity of mining areas. However, in spite of governmental insistence no such attention is being paid resulting in groundwater depletion in the adjoining mining areas. In one of the opencast coal mines in West Bengal, India, groundwater availability in the surrounding villages has deteriorated causing hardship to the villagers. In addition to the normal seasonal fluctuation, water table in the village wells has lowered by 3–4 m. However, the mining authority has been arranging few deep wells in the village areas to supply portable water to the villagers. This can be seen as a short term measure to mitigate the problem. Long term hydrogeological planning is required to be in place in order to replenish groundwater table of the area.

In opencast mines where battery of wells have been sunk to dewater the mines, pumped out water should be supplied to the nearby villages. Such measures have been taken in the Neyveli lignite mines (vide Raghunath H. M., Groundwater).

Wherever, mine water is contaminated due to the effect of mining, chemical analysis of mine water should be done and proper remediation measures as per standard procedures should be taken (vide Younger et al. 2002).

References

Ministry of Environment and Forests, Government of India (2010) Environmental impact assessment guidance manual for mining of minerals, vol 5(2). pp 33–36
Younger PL, Banwart SA, Hedin RS (2002) Mine water: hydrology, pollution, remediation, vol 16. Springer Publishing

Chapter 7
Mine Hydrogeology and Indian Case Studies

Mine hydrogeology deals with the hydrogeological study of a mine, opencast or underground, in order to fulfill various requirements of the mine. Generally, following three requirements arise: (a) to augment water supply to the mines, colonies and process plants, (b) to assess the dimension of ground water seepage in the mine/s and preparation of dewatering plan and (c) to assess both quantitative and qualitative impact of mining on groundwater regime of the area.

For general water supply requirements to the mines and colonies and process control plants, detailed hydrogeological survey needs to be conducted in order to identify potential aquifers. Water well inventory in the area should be conducted and a water table map (iso bath map) or an iso-piezometric (i.e., iso piestic map) map of the area shall be prepared. Hydrogeological investigation is being carried out along with ground water exploration through well drilling.

Pumping tests on the exploratory well in the form of Step Draw down Test (SDT) and Aquifer Performance Test (APT) are to be conducted. These would help to determine the discharge, drawdown and aquifer parameters like co-efficient of Transmissibility, co-efficient of storage and co-efficient of permeability. During pumping test at periodic interval water samples are to be collected for chemical quality of water.

Depending on the water demand of mines, colonies and process plants and based on the pumping test data, planning for ground water development for the area is to be made.

For assessing ground water seepage in the mine, mine development planning has to be understood. The optimum pit depth and pit limit has to be worked out or known from the mine planners and aquifer disposition in the mine area needs to be depicted. If it is an open cast mine, bench height and depth of the mine floor, is required to be known. Basic objective is to keep the water table below the working bench level so that mineral or coal can be excavated without any seepage problem. Therefore, groundwater seepage computation has to be associated with the mine development planning. Bench wise ground water seepage flow can be computed in advance. For any green field mining project, hydrogeological study is a must which can help to plan groundwater seepage control in advance. Bore hole core drilling is done to estimate mineral coal resources. Water level records from all the

P. K. Deb, *An Introduction to Mine Hydrogeology*,
SpringerBriefs in Water Science and Technology,
DOI: 10.1007/978-3-319-02988-7_7, © The Author(s) 2014

boreholes are to be collected and water table map of the area can be constructed. The flow net analysis from the water table map would decide the direction of ground water flow and quantum of ground water flowing through a particular cross section. Identification of potential aquifers can be made from the borehole logs. Aquifers may be sand and sandstone, limestone or Karstic limestone, highly fractured hard rocks like crushed quartzites basalt or granites. Confining materials on top and bottom of the aquifers like clay or shale beds are to be identified which acts as a confining medium and makes the aquifer confined or artesian one.

For a running open cast mine, some times mine gets flooded by inrush of ground water flow hindering mining operation. In such cases, on the pit floor, a sump is made where water is allowed to percolate in it and being pumped out. Interestingly, in most cases, pumped out water returns to the same aquifer causing repeated problem of ground water in rush to the mine. Therefore, thorough hydrogeological study is needed to avoid such problem. Further, mining benches can be identified through which ground water percolation taking place and after a careful hydrogeological study; a battery of dewatering wells may be suggested on the top of the bench so that cumulative pumping is done to keep the water table below the working bench. Pumped out water may be diverted to the nearby stream, if any or to the nearby villages as a part of rural development programme. Hence, care should be taken so that the pumped water does not return to the same aquifer which causes the flooding of the mine. Sometimes, groundwater seepage along steep mining benches causes land slide in the mine. Proper geotechnical studies are to be taken up in such cases to address the problem.

An underground mining project is initiated by driving an incline or adit or by sinking a shaft cutting through the geological formations. This means puncturing of both unconfined and confined aquifers (if any) which contributes towards ground water in rush into the mine. This requires lot of dewatering in the initial phase. Additional water is allowed to be collected in any abundant district where from pumping is carried out for dewatering. Further, working through galleries poses problem of reservoir burst which can be potentially dangerous. Time to time, such big mine accidents do take place claiming many lives of the miners. In India, one such big disaster took place in the 1970s at Chasnala coal mine where more than 400 miners died. That's why for underground mines, preparation of water Danger plan is necessary and statuary requirement.

For a greenfield underground project, the following mine hydrogeological data needs to be collected:

1. Long term annual rainfall data.
2. Hydrogeological study of the area in order to identify potential aquifers and ground water condition and ground water flow direction.
3. Ground water exploration and pumping test data.

Based on these data, a dewatering plan may be prepared for the mining project.

Data collected for mine hydrogeological studies may be used as a computer model for decision making.

Environmental impact assessment study of mining on hydrogeological regime is statuary by the Ministry of Environment and forest of Government of India. Guidance manual for this says that the major impact of deep and large mines (both underground and open cast) is lowering of ground water table which results in reduced groundwater availability. Extraction of different minerals leads to water pollution due to heavy metal, acid discharges and increased suspended solids. However, the impact of mining project on ground water and surface water regime are site specific and depends on the characteristics of the mineral, hydrogeology and requirement of ground water for other uses.

The impact on hydrogeology of the project area may be:

1. Regional surface and ground water movement.
2. Ground water inflow into the mine, with subsequent contact with mining related pollutants.
3. Surface water inflow and precipitation related recharge.
4. Increase in surface and ground water interaction with the mine working because of subsidence.
5. Loss of surface water features such as lakes, streams and ponds through subsidence.
6. Pathways for post closure flow resulting from adits, shafts and overall mine design.
7. Operational and post closure geochemistry and resulting toxics mobility (Ministry of Environment and Forests, Government of India 2010).

Few Indian case studies in respect of mine hydrogeology are being described below:

7.1 Neyveli Lignite Mine, Tamil Nadu

A highly potential confined aquifer (artesian aquifer) occurs in the South Arcot district of Tamil Nadu covering large areas of Cuddalore, Chidambaram and Vrinda Chalam Taluks and is known as Neyveli Aquifer. In the 1960s lot of hydrogeological studies were undertaken in order to establish that the recharge rate is of the order of 90–150 M m^3/annum with a safe average of 120 M m^3/annum. The recharge is mainly due to river flow percolations and rainfall. The recharge area has been computed to be 350–360 km^2. The average annual rainfall in the area is about 1,100 mm and 15–20 % of it trickles down to the sub surface as recharge.

The aquifer may be termed as a coastal aquifer as it occurs adjacent to the sea. The aquifer out crops in a NE–SW direction and it has been estimated that the length along the normal is about 30 km and the average width at right angles to normal is 12 km. The aquifer is exposed in the recharge area which gradually dips in the ESE direction. The top of this confined aquifer occurs at 73.1 m below ground level in the 1st mine and 91.46 m below ground level in the 2nd mine area.

The static water level of the aquifer is more or less steady at 30 m above mean sea level. Therefore, wherever the ground level is below +30 m msl, boreholes drilled give rise to flowing wells. Survey showed that agriculture was drawing about 1,80,000 m³/day of groundwater from the aquifer.

The average hydraulic characteristics of this aquifer are storage co-efficient (S) = 2.5 × 10⁻⁴, co-efficient of permeability (K) = 40 m/day, aquifer thickness (b) = 30 m, co-efficient of transmissibility (T) = Kb = 1,200 m²/day.

The lignite deposit at Neyveli on top of the aquifer separated by a thin layer of clay (thickness varying from 1.5 to 3 m). Before excavating the open cast mine it was estimated that the critical depth was 42.68 m below which it was dangerous to proceed unless the artesian pressure was reduced to below the bottom of the lignite bed (vide Fig. 7.1). Since then dewatering programme started (since 1961) with 28 pumps installed in a rectangular array to give a total discharge of about 114 m³/min. As the mine was depended in order to mine lignite, more water had to be pumped to keep the mine floor from bursting. Peak period pumping was carried out during

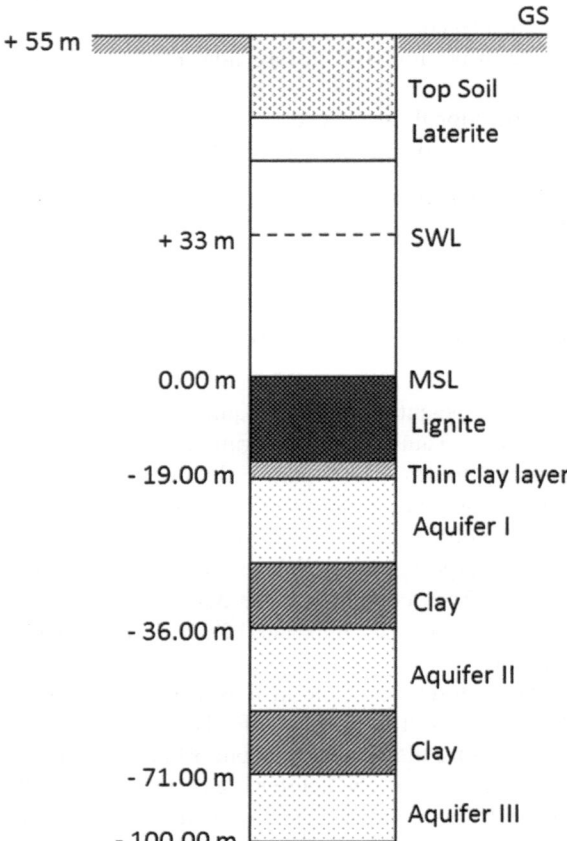

Fig. 7.1 Borelog of Neyveli lighte mine (not to scale)

1962–1964 with the help of 30 more pumping wells in a circular array which were pumped at a rate of 235 m³/min. In order to avoid sea water ingress into the aquifer, in 1969, the pumping wells were installed in the deeper section of the mine in the lignite bench and the pumping was reduced to about 137 m³/min. The Neyveli lignite corporation now pumps at a rate of 1,80,000–2,05,000 m³/day of water and the free flowing agriculture wells account for another 1,60,000–1,80,000 m³/day making a total withdrawal from the aquifer to about 140 M m³/annum. Thus, there balance in respect of annual recharge to the aquifer is being maintained (Raghunath 1987).

7.2 Sonadih Limestone Mine Project, Chattisgarh

Geology of the project area have been described in Chap. 1. The project area falls within part of the Chattisgarh limestone basin which has developed typical Karstic from. Karst is essentially a morphological feature and it implies all the morphological features in limestone terrain. On the surface, there are sink holes small at the surface but often well developed in depth. Within the Sonadih project area a no of pot holes, sinkholes have been observed, various forms of potholes/sinkholes have been observed, some of them have been shown in Fig. 7.2.

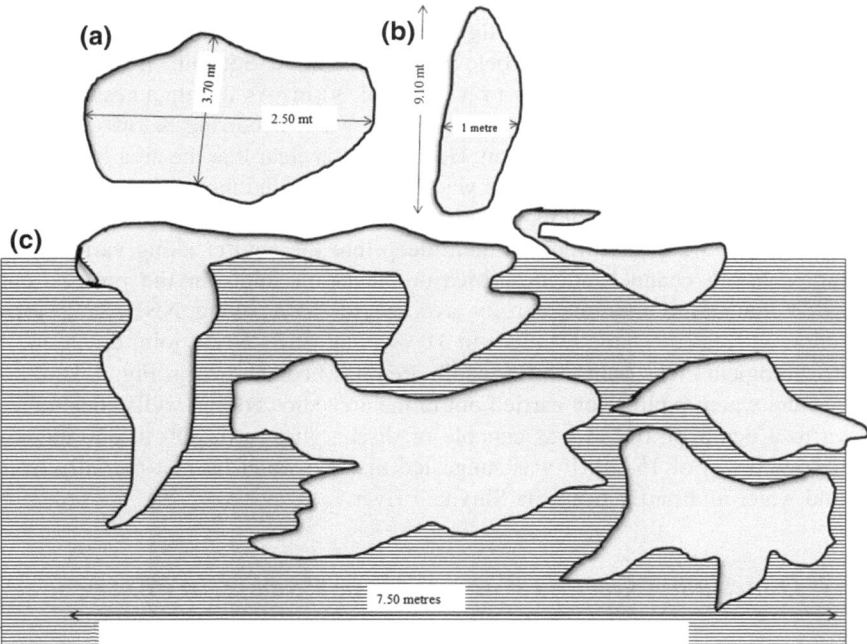

Fig. 7.2 Some geomorphological features of Sonadih limestone project. **a** Sink hole. **b** Swallow hole. **c** Development of complex sinuous trajectories of sink holes and swallow holes

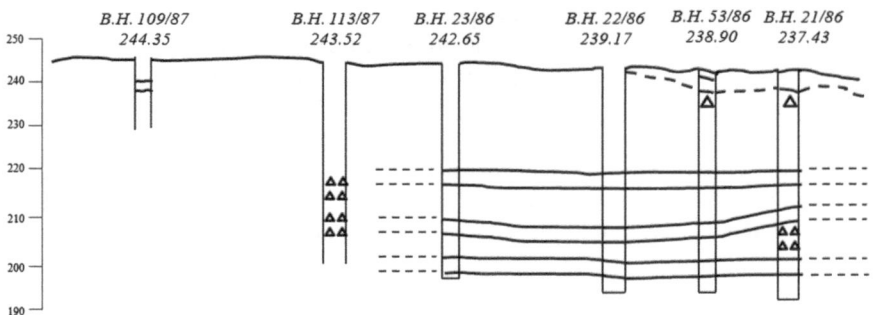

Fig. 7.3 Hydrogeological cross section of Sonadih area

The annual mean rainfall in the area is 982.96 mm. It has been seen that about 15–20 % of this trickles down the surface and joints the ground water body as recharge. Hydrogeology of Karstic limestone differs widely from the alluvium deposits and the sedimentary rocks on the one hand and the igneous and meta-morphic rocks on the other, as it has neither primary porosity of sediments nor the fracture porosity of the effusive rocks. It has thus only secondary porosity and permeability developed essentially as a result of Karstification. Karstification is, therefore responsible for location and movement of both the confined and unconfined ground water bodies. In the unconfined aquifer naturally the first bench of Karstification is one involved below water table. At Sonadih, it appears that unconfined aquifer extends down to a depth of 30 m. As the thickness of shale bands increases below a depth of 30 m, ground water occurring below this depth seems to be under confined condition. However, it is clear that the area offers a set of anisotropic aquifers as the ground water occurrence and movement are entirely controlled by the solution joints. Solution joints opening in the limestone receive rainfall/surface water and transmit them deep into the aquifer along various sec-ondary solution channels/cavities. Measurements on joints in the project area revealed that solution joints density average to 40 % along NNE–SSW joint direction whereas the same averages to 31 % along ESE–WNW joint direction. A hydrogeological cross-section of Sonadih area has been shown in Fig. 7.3.

Ground water exploration carried out in the area showed that wells constructed down to a depth of 100 m was capable of discharging from 300 to 350 m³/day with draw down of 15–20. It was suggested that the water be met partially from ground water and partly from the Shivnath river.

7.2.1 Dimesion of Ground Water Seepage Problem in Mining

Geological exploration carried out till date have revealed the occurrence of cement grade limestone down to an average depth of 16 m. However, at places if occurs up to a depth of 24 and 36 m in patches. It was decided by the mine management to work the deposit by way of open cast mechanized mining with 6 m high benches. Considering above, it was expected that average mine working will be confined to three working benches only i.e., down to a depth of 18 m.

Within this depth, ground water seepage will be there from the top unconfined aquifer. When the 1st bench would be worked, dynamic reserve of the aquifer would account for the ground water seepage. As soon as the second bench is opened, groundwater seepage would take place from the static reserve in addition to the existing seepage from the dynamic reserve (Deb 1987).

Further, during rainy season in addition to ground water seepage, there would be additional percolation from rainfall can be made from the following simple formula:

$$Q_{RF} = S \times Rf$$

where

Q_{RF} Flow due to rainfall within the excavated area in m^3/day

S   Surface area of the excavated mining zone in

and

Rf   Rainfall in meter (to be converted from mm) square meter on a particular rainy day.

A chart indicating therein groundwater seepage and rainfall percolation is shown in the following table (Table 7.1).

7.3 Hydrogeological Investigation for Proposed Cement Plant at Nongkhleih, Jaintia Hills District, Meghalaya

In order to meet the water demand of 1,000 m^3/day for the proposed cement plant (adjacent to the limestone mining area) of a large multi national company, groundwater exploration was carried out in the area. Vide Fig. 7.4.

Geologically, the area is underlain by semi-consolidated to consolidated sedimentary rocks of shella Formation falling under Jaintia Group of Tertiary age. General stratigraphy of the area may be described as under:

Group	Formation	Member
Quarternary	–	Top soil, clay
Jaintia	Shella	Shylhet limestone
		Lower sylhet sandstone (Therria sandstone) with intercalated beds of shale, carbonaceous shale and coal)

Table 7.1 Estimation of groundwater seepage and water percolation due to rainfall in mining

Branch/Level	Water percolation in the pit during lean period					Water percolation in the pit during monsoon period				
	Groundwater seepage from			Percolation due to rainfall (l/min)	Total water percolation (l/min)	Groundwater seepage from			Percolation due to rainfall (l/min)	Total water percolation (l/min)
	Dynamic reserve (l/min)	Static reserve (l/min)	Total groundwater seepage (l/min)			Dynamic reserve (l/min)	Static reserve (l/min)	Total groundwater seepage (l/min)		
1st	753	–	753	Say X	$(753 + X)$	22,200	–	22,200	Say X1	$(22,200 + X1)$
2nd	753	17,300	18,053	Say X	$(18,053 + X)$	22,200	17,300	39,500	Say X1	$(39,500 + X1)$
3rd	753	34,600	35,353	Say X	$(35,353 + X)$	22,200	34,600	56,800	Say X1	$(56,800 + X1)$

Fig. 7.4 Nongkleih site, Meghalaya

Fig. 7.5 Geological map of Nongkhleih area

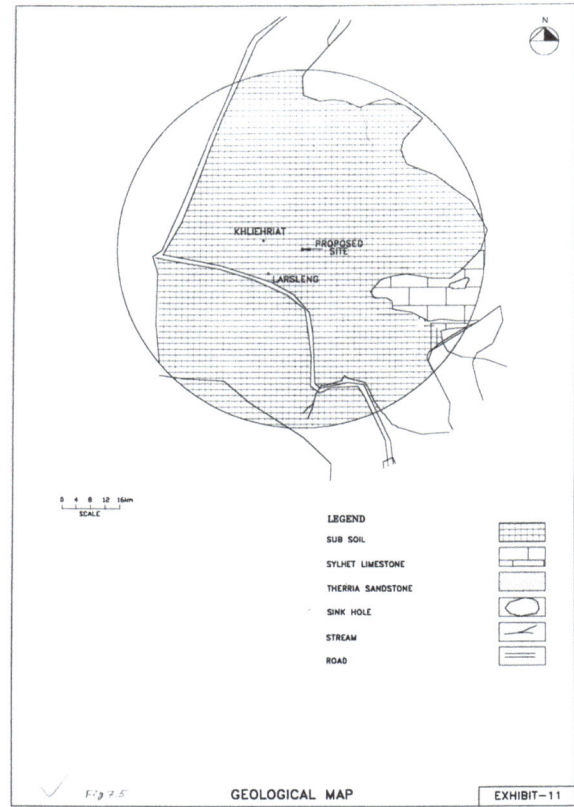

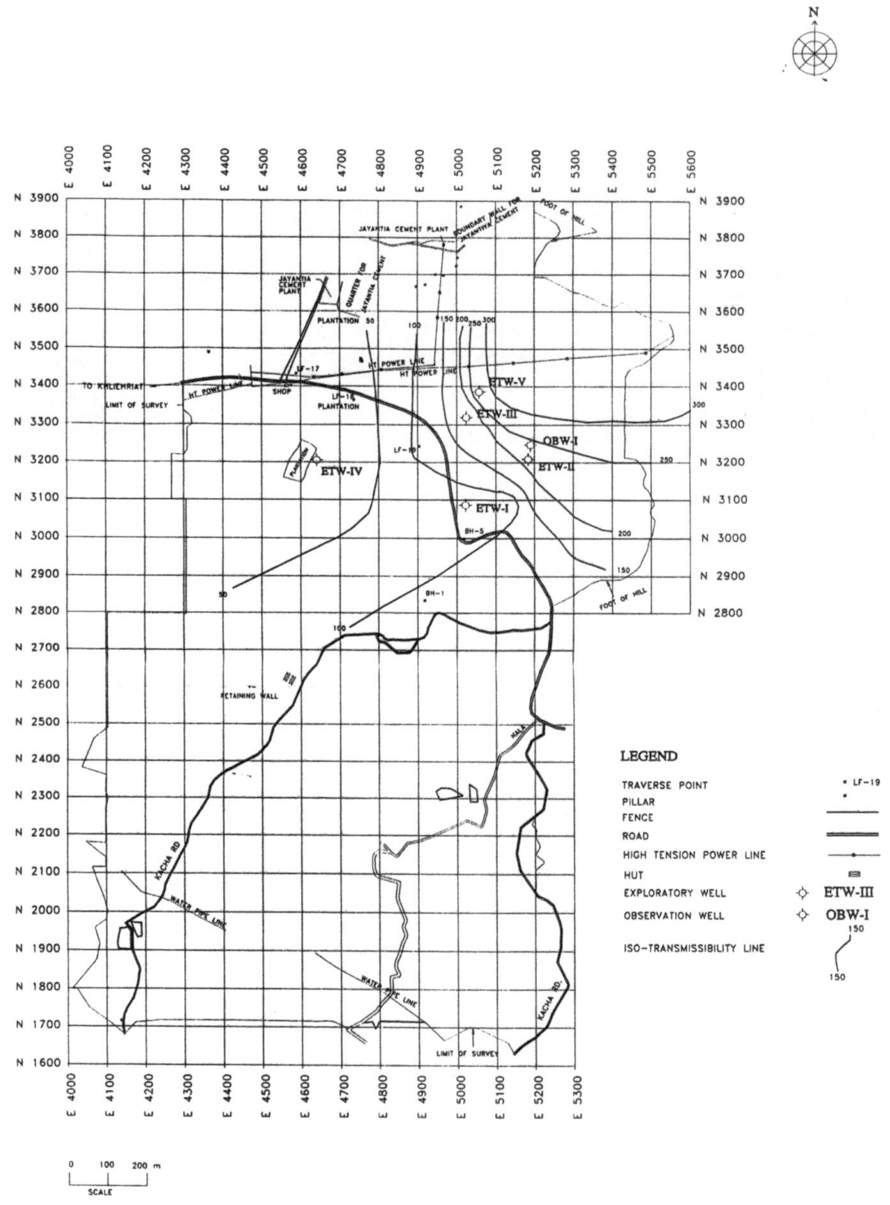

Fig. 7.6 Isopiestic map of Nongkhleih area

The general trend of the rocks of Shella Formation is NE–SW with a variable dip
from 3° to 8° towards SE on the peripheries of the area under study, the limestone is
exposed in the form of a ridge running E–W which later on takes a N–S direction. The

limestone is fossiliferous and bedded in nature. The limestone ridges is at the contact zone of Therria Sandstone and have a few sink holes in them.

The lower Sylhet sandstone (Therria) is overlain by Sylhet limestone. Sandstone is highly weathered and highly fractured and jointed. Thin and impersistent layers of carbonaceous shales, mottled clays and shales are common within this sandstone. The Geological map of the area is shown in Fig. 7.5.

Ground water exploration in the Nongkhleih valley has revealed that the most productive Hydrostratigraphic unit is the Therria Sandstone. However, as the sandstone are mostly semi-conslidated to consolidated in nature, ground water occurrence and movement in them are purely a function of the degree of weathering and secondary porosities like fissures, fractures, joints, lineaments and faults. Exploratory drilling has shown that sandstone aquifers are extensive in both space and time. Five nos exploratory wells and one observation well were drilled and pumping tests were carried out on them. Flow from these wells varied from 13.702 to 64.286 m^3/h with drawdown below 2 m. Co-efficient of transmissibility (T) value worked out as high as 285.33 m^2/day whereas the storage co-efficient (s) was 46.80×10^{-5}.

These values indicated that water demand of the proposed cement plant could be met from ground water resources. As the area was on the ridges, there was no possibility of ground water seepage into the proposed mine/s (Deb 2001). The isopiestic map of the area has been shown in Fig. 7.6.

References

Deb PK (1987) Report on the hydrogeological investigations in the leasehold area of the Cement Plant Project. Tata Steel, Sonadih, Raipur

Deb PK (2001) Hydrogeological investigation for proposed Cement Plant at Nongkleih, Jaintia Hills district, Meghalaya. Final report Lafarge India Limited

Dey AK (1068) Geology of India. National Book Trust of India, New Delhi

Krishnan MS (1968) Geology of India and Burma, vol 7. Higginbothams (P) India Ltd, Madras, p 194

Kruseman GP, De Rider NA (1970) Analysis and evaluation of pumping test data. Institute for Land Reclamation and Improvement, Wageningen, The Netherlands

Ministry of Environment and Forests, Government of India (2010) Environmental impact assessment guidance manual for mining of minerals, vol 5(2). Ministry of Environment and Forests, Government of India, New Delhi, pp 33–34

Raghunath HM (1982, 1987) Groundwater, vol 1. New Age International (P) Ltd, pp 10–13

Theis CV (1933) The relation between the lowering of the piezometric surface and the rate and duration of discharge of a well using groundwater storage. Trans Am Geophys Un 16:519–24

Chapter 8
Conclusion

From a perusal of the foregoing chapters, it may be understood that in case of mine development, the role of mine hydrogeology need not be over-emphasized. When the excavation in a mine intersects the water table of the area, various impacts of mining on hydrogeology of the area are immediately being felt. Groundwater inrush into the mine takes place which necessitates dewatering of the mine by way of pumping. In such cases, in India, random pumping is done to dewater the mine without proper hydrogeological study. It has been emphasized here that groundwater seepage should be properly quantified through systematic hydrogeological studies. Pit limit and pit depth of the mine should be known. Bench height and number of working benches are also required to be known. Thus, if possible, bench-wise quantum of groundwater seepages should be computed in advance, which calls for integration of hydrogeological study with the mine development planning (Deb 1987).

In order to fulfill water demand of the mining colonies and process plant/s, groundwater exploration is carried out through pumping tests (APT) on the exploratory wells so that aquifer parameters like coefficient of Transmissibility (T), Storage Co-efficient (s) and coefficient of permeability can be determined. These aquifer parameters would help in planning for groundwater development. Step Drawdown Test (SDT) is also required to be carried out at various discharges in order to determine optimum yield and optimum drawdown. For groundwater control in the mine, a battery of wells are required to be pumped at a discharge more than optimum yield so that higher drawdown is achieved which would push the water table below the working benches. This would help in uninterrupted mining because the working benches would remain dry.

Mining and its associated activities not only use a lot of water but also affect the hydrological regime of the area. Various impacts of mining on hydrogeological regime has been discussed in the foregoing Chap. 6. Careful analysis of environmental impact on hydrogeology needs to be worked out for a mining project (Ministry of Environment and Forests, Government of India 2010).

P. K. Deb, *An Introduction to Mine Hydrogeology*,
SpringerBriefs in Water Science and Technology,
DOI: 10.1007/978-3-319-02988-7_8, © The Author(s) 2014

References

Deb PK (1987) Report on the hydrogeological investigations in the leasehold area of the cement plant project, Sonadih, Raipur dist, M.P (Phase I), Tata Steel

Ministry of Environment and Forests, Government of India (2010) Environmental impact assessment guidance manual for mining of minerals

Appendix 1
Geological Time Scale

Group/Era	System/Period	Series/Epoch	Age in years from present million (M)
Cenozoic	Quaternary	Holocene/Recent	25,000 years
		Pliesto cene	1 M
	Tertiary	Pliocene	15 M
		Miocene	35 M
		Oligocene	50 M
		Eocene	70 M
Mesozoic	Cretaceous		120 M
	Jurassic		150 M
	Triassic		190 M
	Permian		220 M
	Carboniferous		280 M
Paleozoic	Devonian		320 M
	Silurian		350 M
	Ordovician		400 M
	Cambrian		500 M
Proterozoic	Pre cambrian		1,500 M
Azoic	Achaean		2,000 M

P. K. Deb, *An Introduction to Mine Hydrogeology*,
SpringerBriefs in Water Science and Technology,
DOI: 10.1007/978-3-319-02988-7, © The Author(s) 2014

Appendix 2
Unit Conversion Factors

Length		1 ac ft	= 43,560 cft
1 in.	= 2.54 cm		= 1,233.5 m^3
1 ft	= 0.305 m		= 2.71 × 10^5 imp.gal
1 mile	= 1.609 km		
1 km	= 0.6214 mile	1 ha.m = 10^4 m^3	
			= 8.14 ac ft
Area		*Flow rate (discharge)*	
1 in.^{2}	= 6.452 cm^2	1 cfs (cusec)	= 0.0283 m^3/s (cumec)
1 ft^2	= 0.0929 m^2		= 28.3 lps
1 m^2	= 10.76 ft^2		
	= 1,094 yard^2	1 imp.gpm	= 0.0757 lps
1 ac	= 0.4047 ha		= 1.2 us gpm
	= 4,047 m^2		
	= 43,560 sft	1 us gpm	= 0.063 lps
1 ha	= 10^4 m^2	1 m^3/day	= 2190 imp.gpd
	= 2.471 ac		
1 km^2	= 100 ha	*Permeability*	
	= 0.3861 mile^2	1 darcy = 0.966 × 10^{-2} cm/s	
		11 pd/m^2	= 1.16 × 10^{-6} cm/s
		1 m/day = 1,000 lpd/m^2	
			= 20.44 imp.gpd/ft^2
			= 24.54 us gpd/ft^2
Volume		*Transmissibility*	
1 L	= 1,000 cc	1 m^2/day	= 67.05 imp.gpd/ft
	= 0.22 imp.gal		= 80.52 us gpd/ft
1 barrel = 42 us gal			
1 imp.gal	= 1.2 us gal		
	= 4.546 L		
1 us gal = 3.79 L			
1 m^3	= 1,000 L		
	= 220 imp.gal		
	= 264 us gal		

P. K. Deb, *An Introduction to Mine Hydrogeology*,
SpringerBriefs in Water Science and Technology,
DOI: 10.1007/978-3-319-02988-7, © The Author(s) 2014

About the Author

Pradipta Kumar Deb

Flat 1D, Vikas Apartment, Aswininagar, Baguiati, Kolkata, West Bengal, Pin:-700059, India
Mobile: +91-09830061917; Tel: +91-33-25711113; Email: deb.geosystems@gmail.com

Synopsis

More than 37 years experience in Geological exploration, Mining Geology, Hydrogeology.

Professional Experience

Working as **Advisor (Geology)** in **Black Gold Excavators Pvt. Ltd** for a coal project in Nagaland (**since 2011 till date**).

Worked as **Senior Advisor (Geology)** in **Shyam Group of Industries (2010–2011)**.

Worked as **Deputy General Manager—Geology** in **Shyam Steel Industries Limited (2009–2010)**.

Worked as **Principal Exploration Geologist** in **Salva Resources Pvt. Ltd** (An Australian Consulting Company) and visited South East Asian countries in connection with scooping studies and ore grading quality control of manganese prospect in Mindanao Island of Southern Phillipines. Besides looking after company's business development activity (**2008–2009**).

Worked as **Advisor (Geology)** in **Shyam Group of Industries** and attended to geological inspection of mineral properties both in India and abroad and advising the management on acquisition of mineral rights in iron ore, manganese ore, coal, limestone and base metals. Visited Philippines and widely traveled in the Central Luzon Region and identified laterite nickel, chromite and manganese properties for the company (**2007–2008**).

Worked as **Director (Geology)** in **SRG Services and Consultancy (P) Ltd** and carried out consultancy projects related to geological exploration, mine geology and hydrogeology and guided junior geologists and mining engineers (**2004–2007**).

P. K. Deb, *An Introduction to Mine Hydrogeology*,
SpringerBriefs in Water Science and Technology,
DOI: 10.1007/978-3-319-02988-7, © The Author(s) 2014

Worked as **Head, Geotechnical Division** in **KND Engineering Technologies Ltd** and guided various geotechnical and hydrogeological projects for various companies including Tata Steel (**2002–2004**).

Worked as a **Freelance Consultant** and carried out geological projects like Tin-Tantalum project in Chhattisgarh and ground water exploration projects including Lafarge India Ltd, Meghalaya (**2000–2002**).

Worked as a **Senior Engineering Geologist** in **Fugro-KND Geotech Ltd** and carried out various geotechnical projects including NTPC and Tuticorin Port Trust and ground water pollution project at Rajasthan (**1999–2000**).

Worked as **Senior Consultant** in **GIEM (India) Consortium** and attended to preparation of limestone exploration reports (**1998–1999**).

Worked as a **Consultant** to **Mr. Russi Mody** in **Mobar** and worked on a nickel prospect in Madhya Pradesh and carried out geological appraisal studies in respect of iron and manganese ores in parts of Karnataka. Besides, attended to global trading of metals (**1995–1998**).

Worked as a **Superintending Geologist** in **GIEM (India) Consortium** and supervised many limestone exploration projects for various cement plants (**1994–1995**).

Worked as **Assistant Geologist/Geologist/Assistant Manager/Dy. Manager** in **Tata Steel** and attended to geological exploration of iron, manganese ore, limestone and coal, ore grading and quality control of iron, manganese ores and coal, mine development planning (both short range and long range) and hydrogeological studies of various lease-hold areas of the steel company. Undergone management training at the TISCO Management Development Centre, Jamshedpur and taken up independent charge of people at the mines and laboratory. Contributed significantly in various fields of geology and achieved appreciation from the management (**1979–1994**).

Worked as a **Senior Technical Assistant** (**Hydrogeology**) in **Central Ground Water Board** and carried out systematic hydrogeological investigations and ground water exploration in various parts of North Eastern India (**1975–1979**).

Worked as a **Junior Research Fellow** in the **Department of Geological Sciences of Jadavpur University** (**1974–1975**).